I0472600

Beating Bipolar Disorder

Index:

1.   ## Introduction

Bipolar is one of the most serious
but misunderstood illnesses of our
modern time, with very little facts
known about it, and that which is

known, is at best inadequate and controversial.

There is ongoing speculation as to what causes Bipolar Disorder, so at best, one can only react to the symptoms of the illness, and after thorough diagnosis.

The symptoms of Bipolar Disorder can have a very negative effect on people's lives, jeopardizing their relationships, and putting their work at risk.

In this book, Bipolar Disorder is examined, including the different types of Bipolar, and the known and proved ways of dealing with Bipolar in all its stages.

There are many facets to Bipolar, which some may find confusing, making treatment, especially self-treatment extremely difficult.

I trust that this book will serve its purpose to clear up all misunderstandings regarding Bipolar, and to act as reference concerning the treatment aspects.

2. **Bipolar Disorder Analyzed**

Bipolar disorder is also known as manic depression. It causes serious shifts in mood, energy, thinking, and behavior.

Bipolar Disorder consists of cycles which can last for short periods, or even months.

A person suffering of Bipolar Disorder has to deal with the extreme highs of mania on the one side, and the intense lows of depression on the other.

Mood changes of are so powerful that they interfere with one's overall ability to function.

An example of these two extremes is as follow:

- During a manic episode, one tends to act on impulse, making decisions based on how one feels at that moment. One may do things that are otherwise totally irrational, and without thinking it through. Such person seems also to be very energetic, getting by with the minimum sleep.
- During a depressive episode, the same person would have a very low energy level, be downright depressed over financial matters and debt, with a feeling of utter hopelessness.

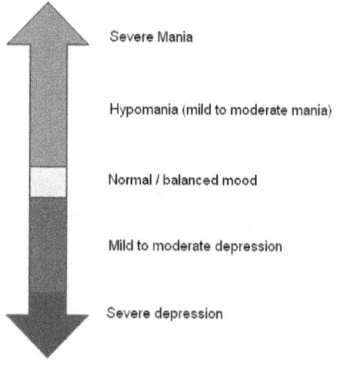

Severe Mania

Hypomania (mild to moderate mania)

Normal / balanced mood

Mild to moderate depression

Severe depression

The first manic or depressive episode of bipolar disorder usually occurs in the teenage years or early adulthood, but because the symptoms can be so subtle and confusing, many people with bipolar disorder are overlooked or misdiagnosed, resulting in unnecessary suffering and delay of treatment.

Although Bipolar Disorder is treatable, a majority of people don't recognize the warning signs of their condition, so they don't get the help they need.

Since bipolar disorder tends to worsen without treatment, it's important to learn what the symptoms look like.

With the proper treatment and support, anyone suffering from Bipolar Disorder can lead a rich and fulfilling life.

3.   Signs and Symptoms

The symptoms vary widely in their pattern, severity, and frequency, and can present itself in different ways in different people, for example:

- Some people experience more manic episodes.
- Some people experience more depression episodes.
- Some people alternate equally between mania and depression.
- Some people have regular mood disturbances.
- Some people experience very little mood disruptions over a lifetime.

There are four types of mood episodes in bipolar disorder:

- Mania,
- Hypomania,
- Depression,

- Mixed episodes.

Each type of Bipolar Disorder mood episode has a unique set of symptoms;

Signs and Symptoms of Mania

Common signs and symptoms of mania include:

- Feeling unusually "high" and optimistic or extremely irritable
- Unrealistic, ostentatious beliefs about one's abilities or powers
- Sleeping very little, but feeling extremely energetic
- Talking so hurriedly that others can't keep up
- Racing thoughts; moving quickly from one idea to the next
- Very distractible and unable to concentrate
- Impaired judgment and impulsiveness

- Acting irresponsibly without thinking about the consequences
- Delusions and hallucinations (in severe cases)

In the manic phase of bipolar disorder, feelings of heightened energy, creativity, and euphoria are common.

People experiencing a manic episode often talk a mile a minute, sleep very little, and are hyperactive. They may also feel like they're all-powerful, invincible, or destined for greatness.

While mania feels good at first, it has a tendency to spiral out of control.

People often behave totally irresponsible during a manic episode: gambling away savings, engaging in inappropriate sexual activity, or making foolish business investments, for example.

They may also become angry, irritable, and aggressive, picking fights, lashing out when others don't go along with their plans, and blaming anyone who criticizes their behavior. Some people even become delusional or start hearing voices.

Signs and Symptoms of Hypomania

Hypomania is not as severe as mania.

People in a hypomanic state feel euphoric, energetic, and productive, but they are able to carry on with their day-to-day lives and they never lose touch with reality.

It may appear to others that people with hypomania are merely in an unusually good mood.

On the other hand, hypomania can result in bad decision making that can potentially harm one's relationship with others, one's career, and reputation.

Hypomania often escalates to full-blown mania or may be followed by a major depressive episode.

Signs and Symptoms of Bipolar Depression

Bipolar depression was dealt with in previous years in the same manner as regular depression, until research indicated to significant differences between the two, particularly when it comes to recommended treatments.

It has become a common fact that most people with bipolar depression are not helped by antidepressants.

There is always a risk that antidepressants can make bipolar disorder worse; triggering mania or hypomania, causing rapid cycling between mood states, or interfering with other mood stabilizing drugs.

There may be some similarities between the two, but certain symptoms are more common in bipolar depression than in regular depression, like irritability, guilt, unpredictable mood swings, and feelings of restlessness.

Some people with bipolar depression also have a tendency to speak and move slowly, sleep a lot, and gain weight. In addition, they are more likely to develop psychotic depression, a condition in which they loose contact with reality.

Common signs and symptoms of Bipolar Depression include:

- Feeling sad, hopeless, or empty inside
- Irritability and temperamental
- Inability to experience happiness
- Tiredness or loss of energy
- Appetite or weight changes
- Mental and physical slowness
- Sleeping too much or too little

- Difficulty to concentrate and memory loss
- Inner feelings of worthlessness or guilt
- Thoughts of death or suicide

Signs and Symptoms of a Mixed Episode

A mixed episode of bipolar disorder features symptoms of both mania or hypomania and depression

Common signs and symptoms of a Mixed Episode include *Depression*, combined with:

- Agitation
- Irritability and temperamental
- Anxiety
- Sleeplessness
- Difficulty to concentrate and memory loss
- Racing thoughts

The combination of high energy and low mood makes for a particularly high risk of suicide.

There are three (3) different types of Bipolar Disorder, namely:

A. *Cyclothymia (hypomania and mild depression)*

Like Bipolar Disorder, Cyclothymia consists of cyclical mood swings.

However, the highs and lows are not severe enough to qualify as either mania or major depression.

To be diagnosed with Cyclothymia, you must experience numerous periods of hypomania and mild depression over at least a two-year time span.

Because people with Cyclothymia are at an increased risk of

developing full-blown bipolar disorder, it is a condition that should be monitored and treated.

B. Bipolar II Disorder (hypomania and depression)

In Bipolar II disorder, the illness involves episodes of severe depression and hypomania, a milder form of mania, but not full-blown manic episodes.

In order to be diagnosed with bipolar II disorder, you must have experienced at least one hypomanic episode and one major depressive episode in your lifetime.

If you ever have a manic episode, your diagnosis would be changed to bipolar I disorder.

C. *Bipolar I Disorder (mania or a mixed episode)*

The classic manic-depressive form of the illness, usually characterized by at least one manic episode or mixed episode, but not always.

The vast majority of people with bipolar I disorder have also experienced at least one episode of major depression, although this isn't required for diagnosis.

4. The causes of BD

There is no singular cause for Bipolar Disorder, since it can be influenced by such a number of factors.

One of those factors might be genetics, in that some people are predisposed to bipolar disorder, but not everyone with an inherited vulnerability develops the illness, strongly indicating that genes are not the only cause.

Brain imaging studies show physical changes in the brains of people with bipolar disorder, while other research points to;

- Neurotransmitter imbalances,
- Abnormal thyroid function,
- Circadian rhythm disturbances,
- High levels of the stress hormone Cortisol.

External environmental and psychological factors are also believed to be involved in the development of bipolar disorder.

These external factors are called triggers.

Triggers can set off new episodes of mania or depression or make existing symptoms worse. However, many bipolar disorder episodes occur without any obvious trigger.

Triggers:

- ### *Stress*

  Stressful events can trigger bipolar disorder in people with a genetic vulnerability. These events tend to involve drastic or sudden changes, either good or bad, such as getting married, going away to college, losing a loved one, getting fired, or moving.

- ### *Substance Abuse*

  While substance abuse doesn't cause bipolar disorder, it can bring on an episode and worsen the course of the disease. Drugs such as cocaine, ecstasy, and amphetamines can trigger mania, while alcohol and

tranquilizers can trigger
depression.

- ***Medication***

  Certain medications, most
  notably antidepressant drugs,
  can trigger mania. Other drugs
  that can cause mania include
  over-the-counter cold medicine,
  appetite suppressants, caffeine,
  corticosteroids, and thyroid
  medication.

- ***Seasonal Changes***

  Episodes of mania and
  depression often follow a
  seasonal pattern. Manic
  episodes are more common
  during the summer, and
  depressive episodes more
  common during the fall, winter,
  and spring.

- **Sleep Deprivation**

  Loss of sleep, even as little as
  skipping a few hours of rest, can
  trigger an episode of mania.

- **Financial Difficulties**

  Worries about finances can
  trigger episodes of Depression.

- **Arguments**

  Arguments also don't cause
  Bipolar Disorder, but getting
  involved in any argument can
  have a very negative effect on
  the condition of a person with
  BP.

5.   Diagnosis of BD

Getting an accurate diagnosis is the
first step in bipolar disorder
treatment.

The mood swings of bipolar disorder can be difficult to distinguish from other problems such as major depression, ADHD, and borderline personality disorder.

Making the diagnosis of bipolar disorder can be tricky even for skilled professionals, so it's best to see a psychiatrist with experience treating bipolar disorder, and having knowledge about the latest research and treatment options.

A diagnostic exam for bipolar disorder generally consists of a Psychological evaluation, Medical history, and a physical examination.

During the evaluation, the Doctor or Bipolar Disorder Specialist will concentrate on the symptoms, the history of the problem, any treatment previously received, and family history of mood disorders.

Screening for thyroid disorders is particularly important, as thyroid problems can cause mood swings that mimic bipolar disorder.

Your doctor or Specialist may also talk to family members and friends about your moods and behaviors.

Often, those close to you can give a more accurate and objective description of your symptoms.

Medical conditions and medications that can mimic the symptoms of bipolar disorder include:

- Thyroid disorders
- Corticosteroids
- Antidepressants
- Adrenal disorders (e.g. Addison's disease, Cushing's syndrome)
- Anti-anxiety drugs
- Drugs for Parkinson's disease

- Vitamin B12 deficiency
- Neurological disorders (e.g. epilepsy, multiple sclerosis)

Bipolar disorder is commonly misdiagnosed as depression. One of the reasons is that most people with bipolar disorder seek help when they're in the depressive stage of the illness.

When they're in the manic stage, they don't recognize the problem, and therefore don't think they need help.

Most people with bipolar disorder are depressed a much greater percentage of the time than they are manic or hypomanic.

Being misdiagnosed with depression is a potentially dangerous problem because the treatment for bipolar depression is different than for regular depression. Antidepressants can actually make bipolar disorder worse.

Indicators that your depression is really bipolar disorder:

- You've experienced recurring episodes of major depression
- You had your first episode of major depression before age 25
- You have a close relative with bipolar disorder
- When you're not depressed, your mood and energy levels are higher than most other people's
- When you're depressed, you tend to sleep too long and overeat.
- Your episodes of major depression are short (less than 3 months)
- You've lost contact with reality while depressed

- You've had postpartum depression before
- You've developed mania or hypomania while taking an antidepressant
- Your antidepressant stopped working after several months
- You've tried 3 or more antidepressants without success.

6. About Treatment

Bipolar disorder requires long-term treatment since it is a chronic illness. One should continue with treatment to prevent relapse.

The most efficient treatment strategy for bipolar disorder involves a combination of:

- Medication
- Therapy

- Lifestyle changes
- Social support.

The recurring and unpredictable manic (ups) and depressive episodes (downs) that characterize the disease make it difficult to lead a stable, productive life.

In the manic phase, you may be hyperactive and irresponsible. In the depressive phase, it may be difficult to do anything at all.

Successful treatment of bipolar disorder depends not only on Medication, but on a combination of factors, namely:

- Self Education on Bipolar Disorder
- Communication with Doctors and Therapists
- Having a strong support system
- Healthy lifestyle choices
- Adhering to the Treatment plan

Know the difference between your symptoms and your true self. Your health care providers can help you separate your true identity from your symptoms by helping you see how your illness affects your behavior.

Be open about behaviors you want to change and set goals for making those changes.

Educate your family and involve them in treatment when possible. They can help you spot symptoms, track behaviors and gain perspective. They can also give encouraging feedback and help you make a plan to cope with any future crises.

Work on healthy lifestyle choices. Recovery is also about a healthy lifestyle, which includes regular sleep, healthy eating, and the avoidance of alcohol, drugs, and risky behavior.

Find the treatment that works for you. Talk to your health care provider about your medications' effects on you, especially the side effects that bother you.

There are many options for you to try. It is very important to talk to your health care provider first before you make any changes to your medication or schedule.

7.   **Treatment Options**

If your doctor determines that you have bipolar disorder, your treatment options will be explained to you. You may also be referred to another mental health professional, such as a psychologist, counselor, or a bipolar disorder specialist.

You will have to work with your healthcare providers to develop a personalized treatment plan, which aims to relieve symptoms, restore your ability to function, fix problems the illness has caused at home and at work, and reduce the likelihood of recurrence. A complete treatment plan involves:

- **_Medication:_** Medication is fundamental for bipolar disorder treatment. Mood stabilizing medication can help minimize the highs and lows of bipolar disorder and keep symptoms under control.

- **_Psychotherapy:_** Therapy is essential for dealing with bipolar disorder and the problems it has caused in your life. Working with a therapist, you can learn how to cope with difficult or uncomfortable feelings, repair your relationships, manage stress, and regulate your mood.

- *Education:* Managing symptoms and preventing complications begins with a thorough knowledge of your illness. Education is a key component of treatment.

- *Lifestyle management:* By carefully regulating your lifestyle, you can keep symptoms and mood episodes to a minimum. This involves maintaining a regular sleep schedule, avoiding alcohol and drugs, following a consistent exercise program, minimizing stress, and keeping your sunlight exposure stable year round.

- *Support:* Living with bipolar disorder can be challenging, and having a solid support system in place can make all the difference in your outlook and motivation. Participating in a bipolar disorder support group gives you the opportunity to share your experiences and learn from others who know what you're going through. The support of friends and family is also invaluable.

8.   **Medication Treatment**

The majority people with bipolar disorder need medication in order to keep their symptoms under control.

Medication continued on a long-term basis can reduce the frequency and severity of bipolar mood episodes, and may prevent them entirely in some instances.

If you have been diagnosed with bipolar disorder, you and your doctor will collaborate to find the right drug or combination of drugs for your specific and individual needs.

Everyone responds to medication differently, so you may have to try several different medications before you find one that relieves your symptoms.

Visit your Doctor regularly

 It's important to have regular blood tests done to make sure that your medication levels are in the therapeutic range. That will enable your Doctor to determine the correct dosage for treatment.

Close monitoring by your doctor will help keep you safe and symptom-free.

Continue taking your Medication

Don't stop taking your medication as soon as you start to feel better. Most people need to take medication long-term in order to avoid relapse.

Don't expect too much from Medication

Some people may expect Medication to cure them completely, and take all their problems away.

Bipolar disorder medication can help reduce the symptoms of mania and depression, but in order to feel your best, it's important to lead a lifestyle that supports wellness.

This includes having people around who supports you, getting therapy, and getting plenty of rest.

Be careful of Antidepressants

Antidepressants can trigger mania or cause rapid cycling between depression and mania in people with bipolar disorder, so it is not particularly effective in the treatment of bipolar depression.

9.   Therapy

Therapy can teach you how to deal with problems your symptoms are causing, including relationship, work, and self-esteem issues. Therapy will also address any other problems you're struggling with, such as substance abuse or anxiety.

Three types of therapy are especially helpful in the treatment of bipolar disorder:

- Cognitive-behavioral therapy

- Interpersonal and social rhythm therapy

- Family-focused therapy

Cognitive Behavioral Therapy

In cognitive-behavioral therapy:

- You examine how your thoughts  affect your emotions.
- You learn how to change negative thinking patterns and behaviors into more positive ways of responding.

For bipolar disorder, the focus is on *managing symptoms*, *avoiding*

triggers for relapse, and **problem-solving.**

<u>Interpersonal Therapy</u>

Interpersonal therapy focuses on current relationship issues and helps you improve the way you relate to the important people in your life, reducing stress, and since stress is a trigger for bipolar disorder, this relationship-oriented approach can help reduce mood cycling.

For bipolar disorder, interpersonal therapy is often combined with social rhythm therapy.

People with bipolar disorder are believed to have overly sensitive biological clocks, the internal timekeepers that regulate circadian rhythms.

This clock is easily thrown off by disruptions in your daily pattern of activity, also known as your "social rhythms." Social rhythm therapy focuses on stabilizing social rhythms such as;

- Sleeping
- Eating
- Exercising.

When these rhythms are stable, the biological rhythms that regulate mood remain stable too.

Family Focused Therapy

Family-focused therapy addresses issues that cause strain in family relationships, and works to restore a healthy and supportive home environment.

A major component of treatment is educating family members about the disease and how to cope with its symptoms, but also focusing on working through problems in the home and improving communication.

10. Alternative Treatment

Most alternative treatments for bipolar disorder are complementary at best, and should be used in conjunction with prescribed medication, and current treatment.

Light and dark therapy

Like social rhythm therapy, light and dark therapy focuses on the sensitive biological clock in people with bipolar disorder. This easily disrupted clock throws off sleep-wake cycles, a disturbance that can trigger symptoms of mania and depression.

Light and dark therapy for bipolar disorder regulates these biological rhythms—and thus reduces mood cycling— by carefully managing your exposure to light.

The major component of this therapy involves creating an environment of regular darkness by restricting artificial light for ten hours every night.

Mindfulness meditation

Research has shown that mindfulness-based cognitive therapy and meditation help fight and prevent depression, anger, agitation, and anxiety.

The mindfulness approach uses meditation, yoga, and breathing exercises to focus awareness on the present moment and break negative thinking patterns.

Acupuncture

Acupuncture is currently being studied as a complementary treatment for bipolar disorder. Some researchers believe that it may help people with bipolar disorder by modulating their stress response.

Studies on acupuncture for depression have shown a reduction in symptoms, and there is increasing evidence that acupuncture may relieve symptoms of mania also.

11.   Recovery Concepts

Obviously a healthy lifestyle is important, meaning that one must make certain adjustments in order to take control of ones life.

The wrong thing to do would be to sit back and think that the medication will do all the work by itself, or that one will get better by ignoring the symptoms.

The following is very important when it comes to things one can do to help oneself on a day-to-day basis;

With good symptom management, it is possible to experience long periods of wellness. Believing that you can cope with your mood disorder is both accurate and essential to recovery.

Depression and manic-depression often follow cyclical patterns. Although you may go through some painful times and it may be difficult to believe things will get better, it is important not to give up hope.

It's up to you to take action to keep your moods stabilized. This includes asking for help from others when you need it, taking your medication as prescribed and keeping appointments with your health care providers.

Become an effective advocate for yourself so you can get the services and treatment you need, and create the life you want for yourself.

Educate yourself about your illness. This allows you to make informed decisions about all aspects of your life and treatment.

Working toward wellness is up to you. However, support from others is essential to maintaining your stability and enhancing the quality of your life.

Diet

Eat plenty of fresh fruits, vegetables, and whole grains and limit your fat and sugar intake. Space your meals out through the day, so your blood sugar never dips too low. High-carbohydrate diets can cause mood crashes, so they should also be avoided. Other mood-damaging foods include chocolate, caffeine, and processed foods.

Get your omega-3s. Omega-3 fatty acids may decrease mood swings in bipolar disorder. Omega-3 is available as a nutritional supplement.

You can also increase your intake of omega-3 by eating cold-water fish such as salmon, halibut, and sardines, soybeans, flaxseeds, canola oil, pumpkin seeds, and walnuts.

12.   Conclusion

Although bipolar disorder tends to be a lifelong, recurrent illness, there are many things you can do to help yourself. You're not powerless. Beyond treatment you get from your doctor or therapist or the medication you take, self-help techniques and simple lifestyle changes can help you manage your moods and stay balanced.

With good coping skills and a solid support system, you can live a full and productive life and keep the symptoms of bipolar disorder in check.

Be actively involved in your own treatment. Learn everything you can about bipolar disorder. Become an expert on the illness.

Study up on the symptoms, so you can recognize them in yourself, and research all your available treatment options. The more informed you are, the better prepared you'll be to deal with symptoms and make good choices for yourself.

Bibliography:

- Depression and Bipolar Support Alliance
- PsychEducation.org